Alcohol, Tobacco and Oral Precancerous Disorders

By

Dr. Manjul Tiwari

*Senior Lecturer (Assistant Professor),
Oral Pathology & Microbiology,
School of Dental Sciences (Hindustan Institute
of Dental Sciences),
Sharda University,
Greater Noida, U.P., India*

River Publishers

Routledge
Taylor & Francis Group
LONDON AND NEW YORK

ISBN: 978-87-92329-85-1

Published, sold and distributed by:
River Publishers
PO Box 1657
Algade 42
9000 Aalborg
Denmark

Tel.: +4536953197
www.riverpublishers.com

*To a large extent, the decrease in
the number of deaths from cancer
of the oral cavity is in the hands
of the dental profession.*
J. C. BLOOD GOOD.

Dedicated To
My Family

Contents

Acknowledgement

Foremost in my mind is my Father Dr. Murli Dhar Tiwari, Director, Indian Institute of Information and Technology, Allahabad, my Mother Dr. Iti Tiwari and my Sister Dr. Maneesha Tiwari about whom it suffices to say—I am fortunate and grateful. For their guidance, supporting, Helping and Guiding me at any time I needed in course of this work.

I would like to extend my heartiest regards and gratefulness to Hon'ble Dr. S.P. Singh, Vice Chancellor and Hon'ble Mr. Pradeep Gupta, Chancellor, Sharda University, India for their kind support and providing facilities to complete the present book.

I would really like to extend my heartiest regards and thanks to Mr. Mohit Tiwari (Brother) and Mrs. Tripti Gautam Tiwari, who supported and helped in every possible way throughout this work and made sure this work is completed as per schedule.

I gratefully acknowledge the help of Mr. Rupesh Tiwari for his statistical jobs.

I am also grateful to Dr. Jagadeesh, Principal, School of Dental Sciences and Dr. Deepak Bhargava, Head of Department, Oral Pathology & Microbiology, School of Dental Sciences, Sharda University for their unconditional help.

There are many others whose names could not be included in this column. That does not mean I am ignoring them; it simply means that they deserve more than my expressions in writing.

Lastly, I vividly acknowledge the one who has made it all possible but perhaps not worthwhile to mention.

Dr. Manjul Tiwari

List of Figures

List of Tables

Abbreviation

S. No.	Word	Abbreviation
1.	Alcohol Drinkers	AD
2.	Bachelor of Dental Surgery	BDS
3.	Betel Quid	BQ
4.	Body Mass Index	BMI
5.	Chewers	C
6.	Confidence Interval	CI
7.	Chewers and Smokers	CS
8.	Dextrene Polystrene Xyelene	DPX
9.	Degree of Freedom	df
10.	Erythroplakia	E
11.	Epithelial Dysplasia	ED
12.	Female	F
13.	Haematoxylin and Eosin	H&E
14.	International agency for Research on Cancer	IARC
15.	Kingdom of Saudi Arabia	KSA
16.	Leukoplakia	L
17.	Mean Sum of Square	M.S.S
18.	Odd's Ratio	OR
19.	Out Patient Department	OPD
20.	Oral Medicine, Diagnosis and Radiology	OMDR
21.	Oral Submucous Fibrosis	OSMF/OSF
22.	Proliferative Verrucous Leukoplakia	PVL
23.	Probability	P
24.	Relative Risk	RR
25.	Smokers	S
26.	Sum of Square	S.S
27.	Socio Economic Status	SES
28.	Smokeless Tobacco	ST
29.	Squamous Cell Carcinoma	SCC
30.	Tobacco with Alcohol users	TA
31.	Verrucous Hyperplasia	VH
32.	Verrucous Lesions	VL
33.	World Health Organization	WHO

1

Introduction

Introduction: In developing countries of South East Asia, Indian people develop more oral precancerous disorders like Leukoplakia, Erythroplakia, Oral submucous fibrosis and Verrucous lesions. The combined effect of chewing tobacco (areca quid chewing) alcohol drinking and smoking greatly increase this risk in oral cavity. In India 75% of cancers of oral cavity are attributable to tobacco chewing, smoking and alcohol drinking.

Aims and objectives: The Purpose of this study is to investigate the prevalence of premalignant disorders, caused in tobacco users, alcohol drinkers and its correlation with the epidemiology, clinical and histopathological features.

Materials and methods: Random collection of samples will be obtained from Moradabad population with or without habits of tobacco using and alcohol drinking.

After that Biopsy were performed from clinically diagnosed lesions like Leukoplakia, Erythroplakia, Oral submucous fibrosis and Verrucous lesions and Histopathological interpretation were will be carried out to correlate the disorder with habits.

Results:

1) The total survey of 35,000 habit positive individuals was done in which 250 patients were identified of different premalignant disorders such as Leukoplakia, OSMF, Erythroplakia, and Verrucous lesions were found.
2) The age group involved in habit positive individual 26–35 and in lesion positive individuals was 36–45 years with male predominance.
3) In clinical positive lesions Leukoplakia was commonest and Erythroplakia was least common.
4) Out of 250 habit positive individuals tobacco chewing was the commonest habit and only 50 patients were readily diagnosed histopathologically.

5) Histopathologically Epithelial dysplasia was commonly found lesion associated with habit. Dysplasia was increased in the lesions as the duration of the habit was increased from 0–5 years to 11–16 years with the presence of carcinomatous changes in the lesion.

Conclusion: From the foresaid study the following conclusions are obtained:

(1) The most commonly affected age was third and fourth decade of life. Males were seen to be more commonly affected than the females.

(2) Premalignant disorders were closely related to the deleterious habits such as tobacco chewing. Tobacco and alcohol drinking together was also considered as a risk factor for such lesions.

(3) As the duration of habits increased, the precancerous disorders became more dysplastic.

Keywords: Oral precancerous disorder, Tobacco, Alcohol, Habits

The concept of oral premalignancy has matured dramatically in recent decades, largely because of the excellence of follow–up studies, the inclusion of clinical parameters and the rejection of the precept that all such lesions must develop into malignancy if left untreated. Most oral precancers are diagnosed on the basis of a combination of clinical and histopathological features.[1] The distinction between a precancerous lesion and a precancerous condition was considered not just academic. At the time these terms were coined, it was considered that the origin of a malignancy in the mouth of a patient known to have a precancerous lesion would correspond with the site of pre-cancer. Oral premalignant lesions are defined as a morphologically altered tissue in which cancer is more likely to occur than in its apparently normal counterpart. Examples of premalignant lesions are leukoplakia and erythroplakia. Precancerous condition are defined as a generalized state associated with a significantly increased risk of cancer. Example Oral Sub mucous fibrosis.[2]

Leukoplakia the most frequent precancerous lesion of the mouth, was first described in the second half of the last century by Hungrian dermatologist, Schwimmer (1877).[3] The word is compounded from the Greek "leucos" meaning white and "plakos" meaning tablet or block. Consequently, it has been used to describe many different kinds of lesion with a white appearance.[4] It is recognized that many different conditions have a superficially similar clinical appearance, that of white plaque and these were formally grouped together under the general designation of leukoplakia.

In this regard, there are claims that this lesion increases the risk of developing oral cancer to more than five times. A high prevalence of oral leukoplakia in any community may therefore be indicative of a strong predisposition to oral cancer. In addition, the use of tobacco has generally been accepted as the principal etiologic factor for oral leukoplakia. Tobacco and Alcohol habits have a strong synergistic or addictive effect.[5]

The term erythroplakia is used analogously to leukoplakia to designate lesions of the oral mucosa that presents as bright red, velvety plaques which cannot be characterized clinically or pathologically as being due to any other condition.[6] Queyrat used the term 'erythroplasie' to designate a red area (plaque) by analogy to the French term 'leucoplasie'. Shear clearly pointed out that if the English word leukoplakia is matched with 'leucoplasie', then Queyrat's 'erythroplasie' should be translated into English 'erythroplakia'.[7] In some cases the red areas of erythroplakia may be intermingled with white patches of leukoplakia.[6] Oral erythroplakia is considered a rare potentially malignant lesion of the oral mucosa. The etiology of oral erythroplakia reveals a strong association with tobacco consumption and the use of alcohol. Transformations rates are considered to be the highest among all precancerous oral lesions and conditions.[7]

Oral Submucous fibrosis first described three decades ago by Pindborg and Sirsat is a chronic, progressive, scarring high–risk precancerous condition of the oral mucosa seen primarily on the Indian subcontinent and in south–east Asia. Describing this condition in five Indian women from Kenya, Schwartz called it 'atrophia idiopathica mucosae oris', subsequently it was called submucous fibrosis by Joshi.[8] It is characterized by excessive production of collagen leading to inelasticity of the oral mucosa and atrophic changes of the epithelium.[9] It is a chronic mucosal condition which predominantly occurs among Indians, people of Indian origin living outside India, occasionally in other Asiatics and sporadically in Europeans. In this conditions the mucosa becomes stiff due to fibroelastic transformation of juxtaepithelial layer leading to inability to open the mouth.[10] The reasons for the rapid increase of the disease are reported to be due to an upsurge in the popularity of commercially prepared areca nut preparations (pan masala) in India and an increased uptake of this habit by young people due to easy access, effective price changes and marketing strategies.[11]

The verrucous lesions described as "an exophytic growth with irregular or blunt projections" included verrucous leukoplakia and verrucous hyperplasia. Pindborg reported that tobacco appears to be a major factor in the causation of verrucous lesions.

Proliferative verrucous leukoplakia (PVL) and verrucous hyperplasia (VH) are two interrelated oral mucosal lesions. Each has been shown to have a considerable propensity to progress to carcinoma either verrucous or conventional squamous cell carcinoma of varying degrees of differentiation. The terms, however, are not clinically or pathologically interchangeable. PVL has no single defining histopathologic lesion and the use of the term is preferably a clinical one. The diagnosis of VH, on the other hand, "must be made Histologically".

The clinicopathologic entity, verrucous carcinoma, was described by Ackerman in 1948 and is sometimes called the Ackerman Tumor. It is a distinct variety of epidermoid carcinoma with pathognomonic clinical appearance, behavior and microscopic features. It occurs most commonly in the oral Cavity. Most writers have cited smoking, snuff-dipping (Steeker, Devine & Harrison 1964) and either chewing tobacco or other types of tobacco products as etiological agents.[12]

Epithelial dysplasia (atypicality, atypia, dyskeratosis) is almost universally regarded as often representing a premalignant condition of the oral mucosa.[13]

In spite of numerous suggested prognostic molecular markers, the presence of epithelial dysplasia as assessed by light microscopic examination is still the strongest predictor of future malignant transformation in an oral potential malignant disorder.[14]

Regardless of individual definitions and classifications, it is generally agreed that:

1. The epithelium with the least proportion of atypical cells has the least risk of being or becoming a carcinoma.
2. The cellular atypia of dysplasia is similar to that seen in squamous cell carcinoma.
3. The final grading or diagnosis should be based on the most severely involved area of change, even if that area involves no more than a few rete processess.[1]

These premalignant disorders are known to be associated with smoking, excess alcohol consumption and areca quid chewing among Asians.[15] Tobacco and undiluted alcohol play a very great role in the development of oral precancerous lesions and oral cancer.

Tobacco use is socially accepted in many segments of Indian society. Tobacco use in India but there are considerable changes in the types and methods by which it is used. According to WHO estimates, 194 million men and 45 million women use tobacco in smoked or smokeless form in India. Only 20% of the tobacco consumed in India by weight is consumed as cigarettes, 40% consumed as bidi and the rest in smokeless form. Tobacco was mainly smokes as bidi, cigarette, hookah. It was used in smokeless form as gul, khaini, gutkha.[16]

The tobacco plant has probably been cultivated by man for many thousands of years although the world–wide distribution and cultivation of the plant did not occur until Spanish and Portugese introduced the plant to the world in the late fifteenth century. Tobacco (picietl) was used by the early American Indians to relieve toothache, to treat skin wounds and insect bites, as an antifatigue agent and as a tooth whitening agent.

The word 'tobacco' is reported to derive from the Spanish Tobago or Tobaca – a term used by the Spanish to describe a Y – shaped instrument used by early American Indians to inhale snuff of various types into the nostrils.

Tobacco is used to manufacture the various forms of smoking tobacco, chewing tobacco and tobacco snuff, derived from two species of the plant genus Nicotiana. The two species are N.tabacum and N. rustica. Tobacco chewing and tobacco snuff usage are generally described as smokeless tobacco.[4] Smokeless tobacco is tobacco that is not burnt when it is used and is usually placed in the oral or nasal cavities against the mucosal sites that permit the absorption of nicotine into the human body. All forms of tobacco use are addictive and cause harm. There are two main types of Smokeless tobacco: chewing tobacco and snuff. It may be used alone or in combination with other substances. Gutkha and pan masala is one of the industrially prepared and highly advertised tobacco products in India and most popular among all smokeless tobacco products especially among youth.[17] Gutkha has been defined as tobacco along with small quantity of pan masala.[18]

Chewing of pan masala was associated with early presentation of OSF as compared to chewing of the betel nut. In recent years, commercial preparation like pan masala have become available in India and abroad.[19] Pan Masala is a Comparatively recent habit in India and is marketed with and without tobacco. Four types of pan masala are available in India market, namely, plain pan masala, sweet pan masala, pan masala containing tobacco, gutkha.[18] Quid has been defined as 'a substance, or mixture of substances, placed in the mouth or chewed and remaining in contact with mucosa, usually containing one or both of the two basic ingredients, tobacco and /or areca nut, in raw or any manufactured or processed form.[20]

Areca nut is the seed of the oriental palm tree areca catechu. The habit of Betel Quid chewing is highly prevalent in India. The composition and method of chewing can vary widely from country to country and the prevalence of Oral Precancerous disorders can vary too. Tobacco is mostly added as a constituents of betel quid in India.[15]

Alcohol drinking appears to increase the risk of oral precancerous disorder by several ways either directly or indirectly. Two types of alcoholic drinks were common: one, fermented extract from palm trees and two, country made liquor which, at times, could be illicit. The third variety available is known as 'Indian made foreign liquor' and carries the standard names such as rum, gin, whisky, etc.[21]

Although a world wide concern regarding the use of tobacco, alcohol and its association with very wide range of oral precancerous disorders is well known. Since previous studies done by various authors **J. J. Pindborg (1963–64),**[22] **P. N. Wahi (1964–66)**[23] reported that the habits of tobacco usage and alcohol drinking was prevalent in Western Uttar Pradesh and Moradabad, a city in north central India, came in the same zone lies on the banks of the Ramganga River, a tributary of the Ganges. Though an important agrarian market, Moradabad is famous worldwide for brass works and glassware, the habit of tobacco and alcohol drinking is found more in this city. So we have conducted the present study as per aims and objectives to find out the possible correlations among the prevalent habits and precancerous disorders.

2

Review of Literature

A premalignant/precancerous phase in the development of oral cancer is predicated by the classic model of experimental carcinogenesis. Virtually all SCC arise from precancerous precursor, but it is difficult to specifically define the term 'Premalignant/Precancerous disorder'. By Precancerous are to be understood changes in the tissue which may assume the character of a malignant tumor at any time, but which, on the other hand, may remain unchanged for a considerable period, particularly if irritations are avoided. These precancerous disorders are associated with the use of deleterious habits such as tobacco and alcohol. Review of literature revealed various interesting observations made by earlier workers in and context of this present study.

In **1946–1976 Jolan Banoczy** done a study on 670 leukoplakia patient in the Department of Maxillo-Facial Surgery and Dentistry and Department of Conservative Dentistry, Semmelweis Medical University Budapest and showed that during a 30-year-period cancer development was found in 40 cases, i.e. 6%. Dysplasia was observed in 24% of the histologically examined leukoplakia with male predominance.[24]

In **1950–74 William G. Shafer** done a study on precancerous disorders to determine the clinical and histologic parameters of the disease and stated that mostly lesions were covered by parakeratin and these lesion were described as leukoplakia while in other lesions non keratinized surface was associated with Erythoplakias.[25]

In **1956–57 Grete Renstrup** conducted a study leukoplakia and establish a diagnosis of leukoplakia in connection with local irritants, such as excessive use of tobacco.[26]

In **1957–73 Jerry E. Bouquot & Robert J. Gorlin** stated that Leukoplakia was the most common of all lesions diagnosed and was the most common of the keratotic lesions, over 35 years of age and was twice as high for males as for females.[27]

In 1960–73 Charles A. Waldron & William G. Shafer stated that Microscopically leukoplakias were varying combinations of hyperorthokeratosis, hyperparakeratosis, and acanthosis without evidence of epithelial dysplasia. Mild to moderate epithelial dysplasia was noted in 12.2% of specimens, and severe epithelial dysplasia or carcinoma in situ was found in 4.5%. Infiltrating squamous cell carcinoma diagnosed in 3.1% of specimens submitted with a clinical diagnosis of leukoplakia.[28]

In 1963–64 J.J. Pindborg et al. done a study on 10,000 patients to record the prevalence of oral leukoplakia and their correlation with the use of tobacco and betel nut and stated that in Lucknow 3.28% were found to have oral leukoplakia and 99.4% of the leukoplakia were found in persons who practiced some form of smoking, chewing or both only 0.6% among those who did not.[22]

In 1964–66 P.N. Wahi done a study in Manpuri District and stated that if the habit of smoking and drinking is added to the habit of chewing Mainpuri tobacco, the risk was found to be highest, being about three times as high among those people with the Mainpuri tobacco chewing habit only. Within the category of those who smoked, the frequency of oral cancer was higher among bidi and cigarette smokers who started chewing at an earlier age, slightly than higher among frequent chewers.[23]

In 1969 Fali S Mehta et al. done a follow-up study of 3,674 Bombay policemen after 10 years and stated that among chewers, the leukoplakia regressed, whereas among smokers, the leukoplakia was found to be more persistent. Although tobacco chewing remained the most prevalent form of tobacco habit, cigarette smoking was found to be on the increase.[29]

In 1976 Jolan Banoczy and **Arpad Csiba**, done a study in order to define the characteristics of epithelial dysplasia in 500 leukoplakias and stated that Epithelial dysplasia was found in 120 cases (24 per cent) and was graded as mild, moderate, or severe.[30]

In 1977 Jens J Pindborg stated that among women, the most frequent habit was reverse smoking (twenty patients) followed by chewing habits (nine patients). In men, the most frequent habit was smoking, either in the ordinary way (fourteen patients) or in the reverse manner (eleven patients). Of the sixty-one lesions showing epithelial dysplasia, thirty-two were present in males and twenty-nine in females. Twenty-one dysplastic lesions were leukoplakias, thirty-one were palatal keratoses (in reverse smokers), and five were cases of submucous fibrosis.[31]

In 1977 Fali S. Mehta et al. done a study in 4000 villagers and showed that reverse smokers exhibiting the various forms of palatal changes. All patients were questioned on smoking and chewing habits.[32]

In 1979–84 Reichart PA stated that considerable differences in the chewing and smoking habits among the various tribes were recorded and some of them were considered tribe-specific. Chewing of betel and miang was more prevalent among older people; these habits seem to have lost their attraction for the younger people. Cigarette smoking was more prevalent among middle-aged individuals. Leukoedema was observed in 12.4%, preleukoplakia in 1.8%, leukoplakia in 1.1% and chewer's mucosa in 13.1%. Men and the older generation were affected more often, except that more women (Karen and Thai) revealed chewer's mucosa.[33]

In 1982 Khin Maung Lay et al. done a survey of 6000 villagers who were above 15 years at the time of study were examined. The prevalence of preleukoplakia was 0.3%, leukoplakia 1.7%, lichen planus 0.4%, leukokeratosis nicotina palati 2.3, erythroplakia 0.1%,submucous fibrosis 0.1%, and cancer 0.03%. The prevalence of submucous fibrosis in their study is lower than in Gujarat (0.2%) and Kerala (0.4%) but is higher than in Andhra (0.04%).[34]

In 1984 Prakash C. Gupta conducted a house-to-house survey in Ernakulam district to study of the association between alcohol habits and oral leukoplakia. The prevalence of leukoplakia was significantly higher among regular (5.7%) and occasional (3.9%) users than among non-users (2.9%) of alcohol. Alcohol usage was found to be related to age as well as tobacco habits. The prevalence of leukoplakia was higher among alcohol with tobacco habit category. The alcohol habit may, perhaps, produce discernible effects only in association with other 'weak' etiological risk factors, such as a single tobacco habit of smoking or chewing rather than a 'strong' etiologic factor such as the mixed habits of chewing and smoking.[21]

In 1985 Murti PR et al., did various population based studies in Ernakulam District, Kerala, India, the cases of Oral Submucous Fibrosis were detected as prevalence and incidence cases in population samples consisting of 27, 600 individuals aged 15 yr and to determined the Malignant transformation rate in oral submucous fibrosis.[35]

John F. Wisniewski and **Alfred A. Bartolucci 1987** did a survey in 858 major league baseball personnel in order to determine the prevalence, cultural distribution and factors influential in, and the level of knowledge of individuals regarding the harmful effects of smokeless tobacco usage.[36]

In the study conducted by **Wolfe MD et al.** in **1987**, 226 Navajo Indians, aged14–19, were interviewed regarding their use of smokeless tobacco (ST), cigarettes, and alcohol. The oral mucosa was examined for evidence of leukoplakia. 64.2% (145) of the subjects (75.4% of the boys and 49.0% of the girls) were users of ST. Of these, over 95% used snuff alone or in combination with chewing tobacco. 55.9% used ST one or more days per week. 52.2% consumed alcohol, usually beer or wine, and 54.0% smoked cigarettes. 25.5% (37) of the users and 3.7% (3) of the non-users had leukoplakia. The duration (in years) and frequency of ST use (days per week) were highly significant risk factors associated with leukoplakia.[37]

In **1988** survey study was carried out regarding identification, socio-economic status, smoking and drinking habits and usage on green yellow vegetables in diet at East godavari district, Andra Pradesh in 78 families by **Satyanarayana gavarasana** and **Maha deva sastri susarla**.[38]

A survery of 599 college students was conducted in **July** and **August 1989** by **Satyanarayana gavarasana et al.** in Andhra Pradesh to formulate an anti smoking policy for youth. In that survey 64.6% were boys and 35.4% were girls between the 15 and 22 years and 8.2% of students (48M + 1F) were smokers.[39]

Prakash C. Gupta in **1989** stated that the incidence of oral cancer is known to be much higher in India than in Northern Europe. Yet both the prevalence and malignant transformation rates of the most important oral precancerous lesion, oral leukoplakia, are reportedly much lower in India than in Northern Europe. Reports from India showing comparatively lower prevalence rates of oral leukoplakia and lower malignant transformation rates than in western countries appear paradoxical because of high incidence rates of oral cancer in India.[40]

Mehta Fali S, et al. in **1989** studied the occurrence of central papillary atrophy of the tongue among tobacco users, its clinical characteristics and the long term behavior in relation to changes in tobacco use in 182 individuals in Ernakulam district, Kerala, India.[41]

Poul Erik Petersen in **1989** describe smoking and alcohol habits of an adult Danish population and to study whether these habits are influenced by living conditions. Moreover, the purpose was to test the hypothesis of unidimensionality of health behavior. He stated that Alcohol consumption as well as smoking was more frequent among workers than officials and smoking and dental health behavior were negatively associated.[42]

Murti PR et al. in **1990**, calculated the Incidence of oral submucous fibrosis from a 10-yr prospective intervention study of 12,212 individuals in

Ernakulam district, Kerala, India with a strong component of health education on tobacco and area nut chewing.[43]

Sinor PN et al. in **1990** reported increase in the relative risk with increase in the duration as well as frequency.[10]

Curtis J. Creath et al., in **1991** stated that higher leukoplakia rates included history of ST use, regular ST use, years of ST use, and the weekly quantity consumed.[44]

In 1992 Tatiana V. Evstifeeva & David G. Zaridze stated that Nass use and cigarette smoking emerged as independent risk factors for oral leukoplakia. In the group with oralleukoplakia, the effect of nass use and cigarette smoking appeared to be additive.[45]

Ying-Chin Ko et al. in **1992** showed that lesser educated older men, blue collar workers, smokers and drinkers were the likeliest betel chewers. A high proportion of chewers was also a smoker and drinker, but tobacco was not found to be chewed together with betel quid.[46]

Maher R et al. in **1994** done a case-control study on chewing and smoking habits and oral submucous fibrosis (OSF) was undertaken in Karachi in 1989/90. Information on habits was collected by personal interview of 157 cases and 157 controls. The male/female risks were found to be similar. Among the cases, an increased risk was observed for areca nut chewing. This habit when practiced alone appeared to have the highest risk followed by pan with or without tobacco.[47]

Reichart, Peter A in **1995** reviewed the Oral cancer and precancer related to betel and miang chewing in Thailand. Chewing of betel and miang as practiced by the Thai was seen mainly in people over 50 years of age. Cigarette smoking was more prevalent among young and middle-aged individuals.[48]

Francis Githua Macigol et al. in **1995** showed that Smoking unprocessed tobacco (Kiraiku) with and smoking cigarettes were the most significant factors. Smoking of both products suggested probable synergistic or additive effects. Commercial beer, wines and spirits were relatively weak, but statistically significant, risk factors. Traditional beer, khat and chillies were not significantly associated with oral leukoplakia.[5]

Ko YC et al. 1995 explore (a) the relationship between the use of betel quid chewing, cigarette smoking, alcohol drinking and oral cancer and (b) synergism between these factors. The synergistic effects of alcohol, tobacco smoke and betel quid in oral cancer were clearly demonstrated, but there was a statistically significant association between oral cancer and betel quid chewing alone. Swallowing betel quid juice (saliva extract of betel quid produced by chewing) or including unripened betel fruit in the quid both seemed to enhance the risks of contracting oral cancer.[49]

Francis Githua Macigo, et al. in **1996** suggest a dose dependent association between oral leukoplakia and the use of tobacco and alcohol in which the number of cigarettes smoked, the quantity of beer consumed. Furthermore. while oral leukoplakia due to cigarette smoking may regress compeletly, those due to kiraiku may persist more than 10 years after cessation of these habits.[50]

T. Nagao et al. in evaluated the validity and outcome of an oral mucosal screening programme conducted **1996–98** in Japan. The oral cavities of 19,056 subjects (31% male, 69% female) were examined by three types of screeners at the Municipal centre of Tokoname city. The screen was recorded as positive for oral cancer or precancer if a mucosal lesion consistent with clinical features of a carcinoma, leukoplakia, erythroplakia or lichen planus was detected.[51]

Tang J-G et al., **(1997)** did Epidemiological survey of oral submucous fibrosis in Xiangtan City, Hunan Province, China. The Yuhu District, one of the five urban districts of the Xiangtan City with a population of 10,0000 was selected as a whole body in the survey. A total of 11,046 individuals were examined; among them were 3907 areca nut chewers (35.37%) and 7,139 non-chewers (64.63%).[52]

Zain RB et al. (**1997**) reported the prevalence of oral mucosal lesions in Malaysia by examining a representative sample of 11,707 subjects aged 25 years and above throughout the 14 states over a period of 5 months during 1993/1994. The prevalence of oral precancer was highest amongst Indians (4.0%) and other Bumiputras (the indigenous people of Sabah and Sarawak) (2.5%) while the lowest prevalence was amongst the Chinese (0.5%).[53]

In **1998 N. Shah et al.** conducted a study to identify the relationship of Oral submucous fibrosis (OSF), with various chewing and smoking habits. Two hundred and thirty-six consecutive cases of OSF were compared with 221 control subjects matched for age, sex and socio-economic conditions in Dental Out-patient's Department at the All India Institute of Medical Sciences, New Delhi. It was found that chewing of areca nut/quid or pan masala (a commercial preparation of areca nuts, lime, catechu and undisclosed colouring, flavouring and sweetening agents) was directly related to OSF.[54]

Allard W F et al. in **1999** explored the possible relationship between a smokeless tobacco preparation (shamma) and oral cancer, among the provinces of the (KSA) Kingdom of Saudi Arabia. Tumor Registry (TR) data from the King Faisal Specialist Hospital and Research Centre (KFSH&RC)

were reviewed for the period from 1976 to 1995. A total of 26,510 Saudi cancer patients were referred over this 20-year period.[55]

Mia Hashibe et al. in **2000** analyzed the effects of chewing tobacco, smoking, alcohol drinking, body mass index, vegetable, fruit, and vitamin/iron intake on the risk of erythroplakia and explored potential interactions between those factors in an Indian population. A more than additive interaction on the risk of erythroplakia was suggested between tobacco chewing and low vegetable intake, whereas a more than multiplicative interaction was indicated between alcohol drinking and low vegetable intake, and between drinking and low fruit intake. Authors concluded that tobacco chewing and alcohol drinking are strong risk factors for erythroplakia in the Indian population.[56]

Jolan Banoczy et al., in **april 2001** gave an overview of the connection between tobacco use and oral leukoplakia, considering the epidemiologic patterns of tobacco habits, the prevalence of smoking in oral leukoplakia and the effect of smoking on clinically healthy oral mucosa. In the data, there is strong evidence that both oral cancer and oral leukoplakia can be induced and promoted by tobacco. Cigarette smoking shows a steep increase in the central European countries, and in these countries the incidence and mortality from oropharyngeal cancer ranks among the highest in the world for both men and women. The proportion of tobacco users (both smoking and smokeless tobacco) among individuals with leukoplakia is high, and a relationship is evident between the tobacco habit and the anatomical location of the leukoplakia. Cross-sectional studies show a higher prevalence of leukoplakia among smokers than among nonsmokers.[57]

A population-based survey was designed by **Yi-Hsin Yang (2001)** to investigate the prevalence of areca/betel quid chewing, oral submucous fibrosis and leukoplakia in a typical aboriginal community of Southern Taiwan. Three hundred and twelve people of 20 years of age or older were collected in the study. The prevalence of chewing areca/betel quid was 69.5%, with an average of 17.3 portions a day for an average 24.4 years. More women (78.7%) than men (60.6%) chewed areca/betel quid. The prevalence of oral submucous fibrosis and leukoplakia were 17.6% and 24.4%, respectively.[58]

Marija Bokor-Bratic et al. 2002 conducted a study to analyze the association between oral leukoplakia and smoking habit, with attention to the duration and quantity of smoking. Of the entire sample of 352 patients aged 40–70 years, 279 were smokers and 73 non-smokers. Oral leukoplakia was found in 53 subjects and among them 50 were smokers and 3 were non-smokers. All smokers had only used cigarettes. The relative risk of developing oral leukoplakia increased with duration of cigarette smoking habit. The majority of smokers with leukoplakia (74.0%) smoked more than 20 cigarettes per day compared to 34.5% of those without leukoplakia. The highest prevalence of leukoplakia (33.3%) was found in subjects who used cigarettes and alcohol. So they concluded that cigarette smoking is significantly related to the etiology of oral leukoplakia.[59]

Rodu B et al., in **2002** examined the prevalence and interaction of cigarette smoking and use of Swedish moist snuff (snus) in the population of Northern Sweden. The cohort study of 2,998 men and 3,092 women aged 25–64 was derived from the Northern Sweden MONICA study, consisting of population based surveys in 1986, 1990, 1994 and 1999. Authors stated that Amongst men ever-tobacco use was stable in all survey years at about 65%, but the prevalence of smoking declined from 23% in 1986 to 14% in 1999, whilst snus use increased from 22% to 30%. In women the prevalence of smoking was more stable in the first three surveys (27%) but was 22% in 1999, when snus use was 6%. In all years men showed higher prevalence of ex-smoking than women. A dominant factor was a history of snus which was more prevalent at younger ages.[60]

Sinha DN et al. in **2003** obtained baseline information about prevalence of tobacco use among school children in eight states in the North-eastern part of India. He stated that the range of current tobacco use (any product) was 63% (Nagaland) to 36.1% (Assam). Current smokeless tobacco use ranged from 49.9% (Nagaland) to 25.3% (Assam). Mizoram reported the highest current smoking (34.5%, mainly cigarette) and Assam reported the lowest (19.7%, again mainly cigarette). Current smoking among girls (8.3% to 28.2%) was also quite high. Over half of current cigarette smokers (53.2% to 96.3%) and a high proportion of current smokeless tobacco users (38.5% to 80.8%) reported feeling like having tobacco first thing in the morning. Only about 20% of students reported having been taught in school about the dangers of tobacco use, except in Mizoram (around 50%). Tobacco use by parents and close friends was positively associated with students current tobacco use. Tobacco use including smoking was very high even among girls, in all eight states in the North-eastern part of India. Signs of tobacco dependency were already visible in these students, more among those who smoked.[61]

M. Hashibe et al. in **2003** conducted a case control study, to examine the association of education, occupation, income and SES index with oral premalignant lesions in Kerala, India. There were a total of 927 oral leukoplakia, 170 oral submucous fibrosis, 100 erythroplakia and 115 multiple oral premalignant lesion cases and 47,773 controls. Subjects with high SES index had protective ORs for oral premalignant lesions, ranging from 0.6 to 0.7, after adjustment for age, sex, BMI, tobacco chewing, smoking, drinking and fruit/vegetable intake. Higher education levels were also associated with decreased risk of all four oral premalignant lesions. Protective ORs for income were observed for oral leukoplakia and possibly oral submucous fibrosis and erythroplakia. SES may be associated with oral premalignant lesions because of access to medical care, health related behaviors, living environment or psychosocial factors.[62]

Urmila Nair et al. in, **2004** stated Gutkha and pan masala have flooded the Indian market as cheap and convenient BQ substitutes and become popular across all age groups wherever this habit is practiced. Chewing of tobacco with lime, BQ with tobacco, BQ without tobacco and areca nut are carcinogenic in humans (IARC, 1985, 2004). These evaluations in conjunction with the available evidence on the BQ substitutes gutkha and pan masala implicates them as potent carcinogenic mixtures that can cause oral cancer. Additionally, these products are addictive and enhance the early appearance of OSF, especially so in young users who could be more susceptible to the disease. Although recently some curbs have been put on the manufacture and sale of these products, urgent action needs be taken to permanently ban gutkha and pan masala, together with the other well-established oral cancer-causing tobacco products. Finally, as the consequences of these habits are significant and likely to intensify in the future, an emphasis on education aimed at reducing or eliminating the use of these products as well as home-made preparations should be accelerated.[63]

S.V. Subramanian, et al. in **2004** investigated the demographic, socioeconomic, and geographical distribution of tobacco consumption in India. He stated that Smoking and chewing tobacco are systematically associated with socioeconomic markers at the individual and household level. Individuals with no education are 2.69 times more likely to smoke and chew tobacco than those with postgraduate education. Households belonging to the lowest fifth of a standard of living index were 2.54 times more likely to consume tobacco than those in the highest fifth. The socioeconomic differences are more marked for smoking than for chewing tobacco. Socioeconomic markers and demographic characteristics of individuals and households do not account fully for the differences at the level of state, district, and village in smoking and chewing tobacco, with state accounting for the bulk of the variation in tobacco consumption.[64]

Sharma Rameshwar et al. in **2004** obtained baseline information about tobacco use prevalence, knowledge and attitude among school personnel in Rajasthan, India. Current daily tobacco use was significantly more among men than women. Four out of nine reported their schools have a tobacco prohibiting policy for both students (48.4%) as well as for school personnel (44.4%) and about same (47.2%) reported their schools enforce its tobacco policy or rule. Over 85% of all school personnel strongly support the tobacco control policies and wanted training in tobacco cessation and prevention.[65]

Y.-H. Yang et al. (2005) had started since 1997 regular follow-up of the study population of a Taiwanese aboriginal community for Incidence rates of oral cancer and oral precancercous lesions. There were 194 persons without any oral lesion in the 1997 screening. During the clinical follow-up investigation and during the analysis of biopsies from precancerous lesions, authors discovered six new lesions (including cancer and precancerous lesions) from five participants. All of the five persons were areca/betel quid chewers, and only one mixed areca/betel quid chewing with cigarette smoking habit. The age-standardized incidence rates for quid lesion, oral submucous fibrosis (OSF) and squamous cell carcinoma (SCC) were 267.0, 374.1 and 146.2 per 10,0000 person-years, for areca/betel quid chewers. So the authors concluded that as compared with the rates from India and the general Taiwanese population, the study community encountered a serious problem of oral lesions.[66]

In the same year he stated that for other oral mucosal lesions, people with mixed habits and chewing only had also significant risks (OR = 8.37 and 3.95, respectively). For both OSF and other oral lesions, the ORs of mixed habits and chewing only were both higher in women than in men. So they concluded that the only way of areca/betel quid could synergize with any tobacco product is through cigarette smoking because the areca/betel quid used in Taiwan does not contain any tobacco product.[67]

Monica A. Fisher et al. 2005 assessed risk factors associated with oral leukoplakia in a US population with high use of smoked tobacco and smokeless tobacco. In their study they stated that Snuff was strongly associated with oral leukoplakia. Individuals currently using smokeless tobacco or snuff were more likely to have oral leukoplakia.[68]

Glorian Sorensen, et al in 2005 assessed social disparities in the prevalence of overall tobacco use, smoking, and smokeless tobacco use in Mumbai, India, by examining occupation, education and gender-specific patterns. Data were derived from a cross-sectional survey conducted between 1992 and 1994 as the baseline for the Mumbai Cohort Study ($n = 81,837$). Odds ratios (ORs) for overall tobacco use according to education level (after adjustment for age and occupation) showed a strong gradient; risks were higher among illiterate participants (male OR = 7.38, female OR = 20.95) than among college educated participants. After age and education had been controlled, odds of tobacco use were also significant according to occupation; unskilled male workers (OR = 1.66), male service workers (OR = 1.32), and unemployed individuals (male OR = 1.84, female OR = 1.95) were more at risk than professionals. The steepest education- and occupation-specific gradients were observed among male bidi smokers and female smokeless tobacco users. The results of this study indicate that education and occupation have important simultaneous and independent relationships with tobacco use that require attention from policymakers and researchers alike.[69]

Jyotsna Changrani, et al. in 2006 stated that Indian-Gujaratis Immigrants in New York City are more likely to identify gutkha as causing oral cancer. Between the two communities, there were significant differences in paan and gutkha usage, migration effects, and oral cancer risk perception.[70]

Marcio Diniz Freitas et al. 2006 did a Retrospective study of 52 patients with oral leukoplakia. Clinical and pathologic data (age, sex, lesion size, lesion location, and presence/absence of dysplasia) were compared between 41 current-smoking patients and 11 never-smoking patients. They stated that nonsmoking-related oral leukoplakia lesions are more frequent among women than among men, are more likely to be located on the tongue than smoking-related lesions, and show epithelial dysplasia more frequently than smoking-related lesions. The mean age of the smoking patients was 49 yrs, significantly lower than the never-smoking patients (59 yrs)($P < 0.05$). The proportion of women was markedly and significantly higher in the never-smoker group than in the smoker group (82% vs. 22%). The odds ratio for lesions on the tongue (0.80, 95% CI 0.01–0.37) was statistically significant at the 5% level (i.e., 95% CI). Dysplastic lesions were observed most frequently in the never-smoking patients, and this difference was statistically significant ($P < .026$).[71]

Sardar Z Imam, et al. in **2007** conducted a study with the objective of determining the prevalence of smokeless tobacco among Pakistani medical students. A cross sectional study was carried out in three medical colleges of Pakistan–one from the north and two from the southern region. 1,025 students selected by convenient sampling completed a peer reviewed, pre-tested, self-administered questionnaire. Two hundred and twenty (21.5%) students had used tobacco in some form (smoked or smokeless) in their lifetime. Sixty six (6.4%) students were lifetime users of smokeless tobacco.Thirteen (1.3%) were daily users while 18 (1.8%) fulfilled the criterion for established users.

Naswar was the most commonly used form of smokeless tobacco followed by paan and nass. On univariate analysis, life time use of smokeless tobacco showed significant associations with the use of cigarettes, student gender

($M > F$), student residence (boarders > day scholars). Multivariate analysis showed independent association of lifetime use of smokeless tobacco with concomitant cigarette smoking, student gender and location of the medical college.[72]

A hospital-based cross-sectional study on various habit patterns associated with OSF was performed in Nagpur over a 5-year period by **V. K. Hazarey et al. in 2007**. A total of 1000 OSF cases from 26,6418 out patients comprised the study sample. The male-to-female ratio of OSF was 4.9:1. Occurrence of OSF was at a significant younger age group (< 30 years) among men when compared with women. Exclusive areca nut chewing habit was significantly more prevalent in women. Whereas significant increase for Gutkha (Areca quid with tobacco) and kharra/Mawa (crude combination of areca nut and tobacco) chewing was found in men when compared with women. They conclude that there is a marked difference in literacy, socioeconomic status, areca nut chewing habits, symptoms and disease severity in women when compared with men in the central Indian population.[73]

3

Results

All the patients reported to OPD of Kothiwal Dental College and Research Centre in a period of 1st January 2007 to 30th July 2008 were questioned for the type and nature of oral habits by the BDS final year students posted in the Department of Oral Medicine irrespective of chief complaint. The habit positive patients were segregated and further examined for the presence of any oral lesions by posted interns in the same department. After that the clinically positive premalignant disorder cases were further examined by me under the supervision of faculty members in Department of Oral Pathology and Microbiology. After taking written consent from the patients the biopsy was performed for histopathological evaluation. The tissue was processed, sections obtained were stained with H&E and examined. The findings were analyzed & obtained result is presented hence forth:

The present survey study was conducted from 1st January 2007 to 30th July 2008 in Moradabad population to find out the premalignant disorder in Tobacco users and Alcohol drinkers with special emphasis given on Epidemiology, Clinical and histopathological aspects. The study was done in Out Patient Department (OPD) of Kothiwal Dental College and Research Centre, Moradabad and Oral Cancer Detection Camps held in and around Moradabad.

The total survey of 35,000 habit positive individuals was done during this period. Out of 35,000 habit positive individuals 28,430 individuals were selected from OPD of Kothiwal Dental College and Research Centre, Moradabad and remaining 6,570 were from 8 Oral Cancer Detection Camps during this period. In 35,000 habit positive individuals 34,639 were males and 361 were females (Fig. 3.1).

Habits were divided into Chewers (C), Smoker (S), Chewers + Smoker (CS), Alcohol drinkers (AD) and Tobacco with Alcohol users (TA). Out of 35,000 individuals it was found that 8,824 individuals were chewers (25.21%), 6,854 were smokers (19.58%), 7,534 were chewers as well as

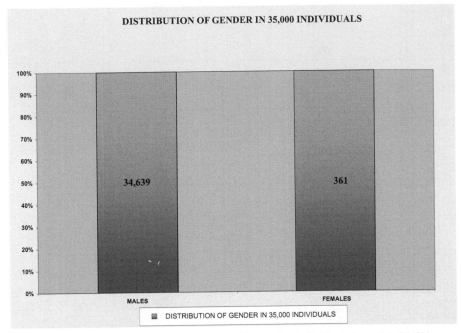

Figure 3.1 Distribution of gender in 35,000 habit positive individuals in which 34,639 were males and 361 were females.

smoker (21.53%), 748 were alcohol drinkers (2.14%) [Fig 3.10] and 11,040 were tobacco users with alcohol drinkers (31,545) (Fig. 3.2).

The age groups with gender distribution and habits were also recorded in 35,000 individuals. In 35,000 individuals age groups i.e. 16–25 years, 26–35 years, 36–45 years 46–55 years, 56–65 years, above 65 years were noted [Table 3.1]. Out of 35,000 individuals 4,723 were from 16–25 year of age groups in which 4,630 were males (98.03%) and 93 were females (1.97%). Amongst 4,630 males 2,166 were C (46.78%), 846 were S (18.27%), 962 were CS (20.78%), 71 were AD (1.533%) and 585 were TA (12.64%) and in 93 females 82 were C (88.17%) and 11 were S (11.83%).

Between the age groups of 26–35 years total 10,972 were examined out of which 10,824 were males (98.65%) and 148 were females (1.35%). Amongst 10,824 3,042 was C (28.104%), 3,279 was S (30.29%), 925 was CS (8.55%), 240 AD (2.22%) and 3,338 TA (30.84%) and in 148 females 104 were C (70.27%) and 44 were S (29.73%).

In age groups 36–45 9,622 individuals 9,569 were males (99.45%) and 53 females (0.55%). Amongst 9,622, 1,863 were C (19.47%), 1,375 were S

Table 3.1 Percentage of Age, Gender and Habit in 35,000 Individuals

Age, Gender and Habit in 35,000

Age Groups (in years) (Total no.of user out of 35,000)	Gender (M = Male, F = Female) Chewers	Smokers	Chewer+ Smoker	Habit — Alcohol	Habit — Tobacco+ Alcohol	
16–25 (4723)	M = 4630 (**98.03%**)	2166 (**46.78%**)	846 (**18.27%**)	962 (**20.78%**)	71 (**1.533%**)	585 (**12.64%**)
	F = 93 (**1.97%**)	82 (**88.17%**)	11 (**11.83%**)			
26–35 (10,972)	M = 10,824 (**98.65%**)	3042 (**28.104%**)	3279 (**30.29%**)	925 (**8.55%**)	240 (**2.22%**)	3338 (**30.84%**)
	F = 148 (**1.35%**)	104 (**70.27%**)	44 (**29.73%**)			
36–45 (9622)	M = 9569 (**99.45%**)	1863 (**19.47%**)	1375 (**14.37%**)	3079 (**32.18%**)	92 (**96.14%**)	3160 (**33.02%**)
	F = 53 (**0.55%**)	25 (**47.17%**)	28 (**52.83%**)			
46–55 (5786)	M = 5744 (**99.27%**)	724 (**12.61%**)	593 (**10.33%**)	1353 (**23.56%**)	119 (**2.07%**)	2955 (**51.45%**)
	F = 42 (**0.7254%**)	24 (**57.14%**)	18 (**42.86%**)			
56–65 (2897)	M = 2881 (**99.45%**)	666 (**23.12%**)	509 (**17.67%**)	854 (**29.64%**)	145 (**5.032%**)	707 (**24.54%**)
	F = 16 (**0.5523%**)	7 (**43.75%**)	9 (**56.25%**)			
Above 65 (1000)	M = 991 (**99.1%**)	119 (**12.01%**)	135 (**13.63%**)	361 (**36.43%**)	81 (**8.17%**)	295 (**27.77%**)
	F = 9 (**0.9%**)	2 (**22.22%**)	7 (**77.78%**)			

PREVALENCE OF HABITS IN 35,000 INDIVIDUALS

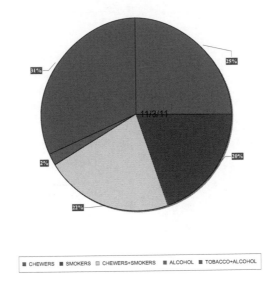

Figure 3.2 The prevalence of habits (chewer, smoker, chewers + smoker, alcohol drinker and tobacco + alcohol users)in 35,000 individual was 25.21%, 19.58%, 21.53%. 2.14% and 31.54% found. The prevalence of tobacco = alcohol users in 35,000 individual was found high i.e 31.54%.

(14.37%), 3,079 were CS (32.18%), 92 were AD (96.14%) and 3,160 were TA (33.02%) and in 53 females 25 were C (47.17%) and 28 were S (52.83%). In 46–55 age group total of 5,786 individuals 5,744 were males (99.27%) & 42 females (0.7254%). Amongst 5,744 males 724 were C (12.61%), 593 were S (10.33%), 1,353 were CS (23.56%), 119 were AD (2.07%) and 2,955 were TA (51.45%). Out of 42 females 24 were C (57.14%) and 18 were S (42.86%).

In 56–65 age group total of 2,897 individuals 2,881 were males (99.45%) and 16 were females (0.5523%). Amongst 2,881 males 666 were C (23.12%),

Table 3.2 Duration of Habits in 35,000 Individuals

	Chewers	Smokers	Chewers + Smokers	Alcohol	Tobacco + Alcohol
0–5 YEARS	2930	1416		1894	
6–10 YEARS	9003	4881		1551	
11–15 YEARS	7080	3662	425		1118
16–20 YEARS			350		690

509 were S (17.67%), 854 were CS (29.64%), 145 were AD (5.032%) and 707 were TA (24.54%). Out of 16 females 7 were C (43.75%) and 9 were S (56.25%). In age group of above 65 total 1,000 individuals were examined out of them 991 were males (99.1%) & 9 were females (0.9%). Amongst 991 males 119 were C (12.01%), 135 were S (13.63%), 361 were CS (36.43%), 81 were AD (8.17%) and 295 were TA (27.77%). In 9 females 2 were C (22.22%) and 7 were S (77.78%). (Table 3.2).

Amongst 35,000 habit positive individual Oral lesion (Leukoplakia, OSMF, Erythroplakia and Verrucous lesion) were seen in 250 patients. Amongst 250 patients 241 patients were males and 9 patients were females. 223 patients were identified from OPD of Kothiwal Dental College and Research Centre, Moradabad and 27 patients from Oral Cancer Detection camps (Fig 3.3, 3.4, 3.5, 3.6).

In 250 patients the 214 patients were of leukoplakia (85.60%), 32 patients were of OSMF (12.80%), 3 patients were of Verrucous lesion (1.20%), 1 patient with erythroplakia (0.40%) and number of habits were C 164 (65.60%), S 64 (25.60%), CS 3 (1.20%), AD 11 (4.40%) and TA 8 (3.20%) (Fig. 3.7, 3.8).

The prevalence rate of premalignant disorder was also establish out of total sample size as follows Leukoplakia (214) was 0.6114%, OSMF (32) 0.0914%, Verrucous lesion (3) 0.0086% and Erythroplakia (1) 0.0029% found (Fig. 3.9).

The age groups, gender, and their relation with premalignant disorders were also recorded. In age group between 16–25 years 34 male and 3 females patients (37) were identified out of which leukoplakia was found in 35 patients (male 32 and female 3) and OSMF in 2 patients. 48 patients (46 males and 2 females) were found between the Age group of 26–35 years out of them leukoplakia was found in 40 patients (38 males and 2 females) and OSMF in 8 patients. Between the age group of 36–45 years 83 patients (male 81 and female 2) were found. Out of 83 patients leukoplakia found in 73 (71 males

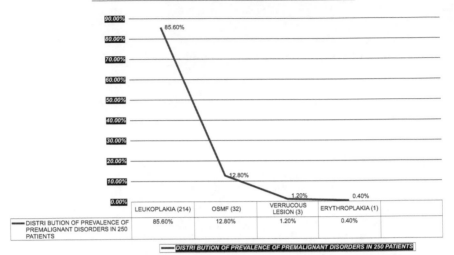

Figure 3.3 Out of 35,000 individual which was all habit positive 250 patients were those in which premalignant disorder were found. In 250 patients 214 patient had Leukoplakia, 32 had OSMF, 3 had Verrucous lesion and only 1 patient had Erythroplakia. Among these premalignant disorders the prevalence rate of Leukoplakia was found high. The prevalence rate of premalignant disorder in 250 patient was Leukoplakia (85.6%), OSMF (12.8%), Verrucous lesion (1.2%) and Erythroplakia (0.4%).

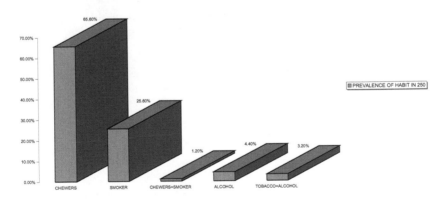

Figure 3.4 The prevalence rate of habit out of 250 individual in which premalignant disorder (Leukoplakia, OSMF, Erythroplakia and Verrucous lesion) was identified positive is 65.6% in chewer highest among other habits like smoker 25.6%, chewer + smoker 1.20%, alcohol 4.40% and alcohol + tobacco users 3.20%.

PREVALENCE OF PREMALIGNANT DISORDER (250) OUT OF 35,000

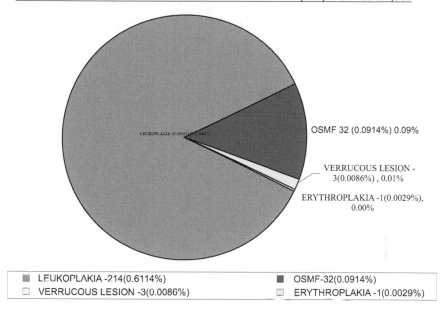

OSMF 32 (0.0914%) 0.09%

VERRUCOUS LESION - 3(0.0086%) , 0.01%

ERYTHROPLAKIA -1(0.0029%), 0.00%

■ LEUKOPLAKIA -214(0.6114%)	■ OSMF-32(0.0914%)
☐ VERRUCOUS LESION -3(0.0086%)	☐ ERYTHROPLAKIA -1(0.0029%)

Figure 3.5 The prevalence of premalignant disorder in 250 patients was also compared with out of 35,000 individual then the prevalence rate of Leukoplakia was 0.6114%, OSMF 0.0914%, Verrucous lesion 0.0086% and Erythroplakia 0.0029% found.

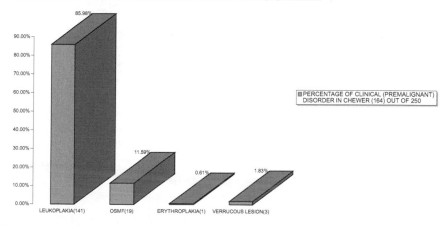

Figure 3.6 In 250 patients the number of chewers was 164. Percentage of premalignant disorders in chewers was calculated. In chewers leukoplakia (141) 85.98%, OSMF (19) 11.59%. Verrucous lesion (3) 1.83% and Erythroplakia (1) 0.6098% was found.

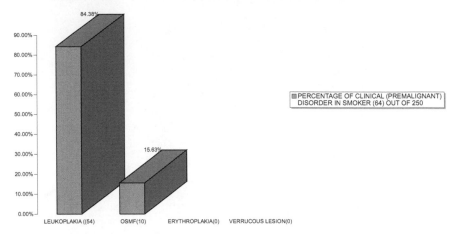

PERCENTAGE OF CLINICAL (PREMALIGNANT) DISORDER IN SMOKER (64) OUT OF 250

Figure 3.7 In smokers (64) leukoplakia (54) 84.38% and OSMF (10) 15.63%.

Table 3.3 Mean of Age Group

Age Group (in years)	No.of Patients (out of 250)	Mean of Age Group	Percentage
16–25	37	20.5	758.5
26–35	48	30.5	1464.0
36–45	83	40.5	3361.5
46–55	69	50.5	3484.5
56–65	10	60.5	605.0
Above 65	3	0	195
Total	**250**	—	**9868.5 = 39.47%**

and 2 females), OSMF in 9, Erythroplakia in 1 patient. From 46–55 years of age groups 69 patients (68 males, 1 female) out of them leukoplakia was found in 60 patients (59 males, 1 female), OSMF in 9 patients. 10 (10 males) patients were found in age group of 56–65 years out of them leukoplakia was found in 6 patients and OSMF in 3 patients. 3 (male 2, female 1) patients of above 65 years were found in which verrucous lesion found in 2 patients (male 1, female 1) and OSMF found in 1 patient (Table 3.3).

By applying **One way Anova F and paired Z test** Correlation between Duration & Type of habit and their association in variation of Oral lesions were also establish. A significant difference was found among the clinical

PERCENTAGE OF CLINICAL (PREMALIGNANT) DISORDER IN CHEWER+SMOKER (3) OUT OF 250

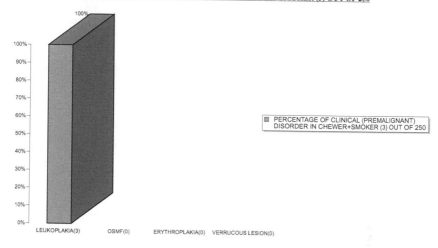

Figure 3.8 In Chewers as well as smokers (3) all 3 patient had leukoplakia 100%.

PERCENTAGE OF CLINICAL (PREMALIGNANT) DISORDER IN TOBACCO+ALCOHOL (8) OUT OF 250

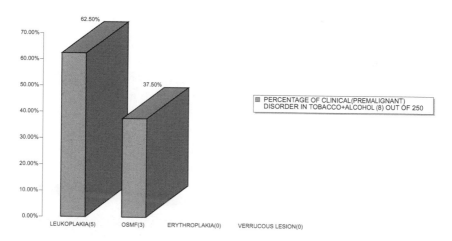

Figure 3.9 In (8) tobacco users with alcohol leukoplakia (5) 62.5% and OSMF (3) 37.5% was found.

PERCENTAGE OF CLINICAL (PREMALIGNANT) DISORDER IN ALCOHOL(11) OUT OF 250

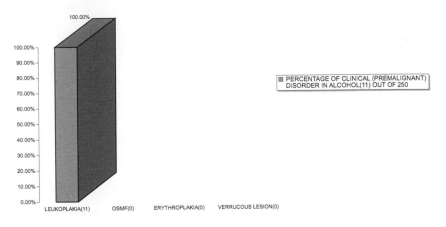

Figure 3.10 In alcohol drinkers (11) 11 patients were found all had leukoplakia 100%.

lesions at 5% level of significance with duration of time. A significant difference was observed for C and S i.e **P < 0.05%**. A significant difference was observed from 11–15 and 16–20 years of duration i.e **P < 0.05%**. In C between 0–5 years duration 67 patient with leukoplakia, 5 patient with OSMF were found. In 6–10 years duration 60 patients with leukoplakia, 6 with OSMF were found. In 11–15 years 14 Leukoplakia, 6 OSMF, 1 Erythroplakia and 3 verrucous lesion were found. In 16–20 years only 2 patient of OSMF were found. In S 10 patient of leukoplakia were found in between the duration of 0–5 years. In 6–10 years 17 patients of leukoplakia and 1 patient of OSMF were found. In 11–15 years 27 patients of leukoplakia and 9 patients of OSMF was identified. In 16–20 years no patients were found. In CS only 3 patient of leukoplakia were found between the duration of 11–15 years and 16–20 years. In TA 5 patients of leukoplakia between 16–20 years of duration and 3 patients of OSMF out of which 1 with 11–15 years duration and 2 with 16–20 years of duration. In AD 7 patient of leukoplakia with 0–5 years of duration and 4 patient of leukoplakia with duration of 6–10 years (Table 3.4).

Out of 250 only 50 patients consented or agreed for biopsy. Amongst 50 patients 39 patients were C (78.0%), 5 S (10.0%), 1 was CS (2.0%) and 5 patient were TA (10.0%) (Fig 3.10, 3.11, 3.12, 3.13, 3.14).

The histopathological findings were further correlated with type & duration of habit and clinical presentation of lesion.

Table 3.4 Correlation of Habit, Clinical Lesion and Duration of Habit in 250 Patients

Habit	Duration (in years)	Leukoplakia	OSMF	Erythroplakia	Verrucous Lesion
Chewers					
	0–5	67	5	0	0
	6–10	60	6	0	0
	11–15	14	6	1	3
	16–20	0	2	0	0
Total		**141**	**19**	**1**	**3**
Smokers					
	0–5	10	0	0	0
	6–10	17	1	0	0
	11–15	27	9	0	0
	16–20	0	0	0	0
Total		**54**	**10**	**0**	**0**
Chewers + Smokers	0–5	0	0	0	0
	6–10	0	0	0	0
	11–15	2	0	0	0
	16–20	1	0	0	0
Total		**3**	**0**	**0**	**0**
Tobacco + Alcohol	0–5	0	0	0	0
	6–10	0	0	0	0
	11–15	0	1	0	0
	16–20	5	2	0	0
Total		**5**	**3**	**0**	**0**
Alcohol	0–5	7	0	0	0
	6–10	4	0	0	0
	11–15	0	0	0	0
	16–20	0	0	0	0
Total		**11**	**0**	**0**	**0**

In C with 0–5 years of duration 67 out of 250 patients were found in which histopathological correlation of 11 patients (16.42%) were noted. Out of 11 patients 6 were mild epithelial dysplasia (54.55%), 4 moderate epithelial dysplasia (66.67%) and 1 with hyperorthrokeratinized epithelium (9.10%) were diagnosed. But with 6–10 years of duration 60 patients of leukoplakia in which histopathological correlation of 9 patients (15%) was noted. Out of 9 patients 1 was mild epithelial dysplasia (11.11%), 5 moderate epithelial dysplasia (55.56%), 1 severe epithelial dysplasia (11.11%), 1 carcinoma in situ (11.11%), and 1 with atrophic epithelium with inflamed connective tissue (11.115) was diagnosed. In 14 patients of leukoplakia with 11–15 years of duration only 1 carcinoma in situ (7.142%) was diagnosed.

In S with 0–5 years of duration 10 patients of leukoplakia were evaluated out of them 1 with epithelial hyperplasia (50%) and 1 with hyperorthrokeratinization (50%) (total 2 (14.28%)) were diagnosed. In 6–10 years duration

PREVALENCE OF HABIT IN 50 PATIENTS(HISTOPATHOLOGICAL EVALUTION)

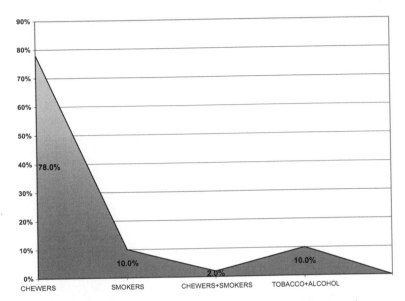

Figure 3.11 Out of 250 individual 50 patients with habit i.e. chewers, smoker, chewer + smoker and alcohol + tobacco user agreed for biopsy in which the prevalence rate was 78%, 10%, 2% and 10%. None of alcohol user was agreed for biopsy.

1 with acanthosis (50%) and 1 with hyperparakeratinzation with acanthosis (50%) (2 (11.76%)) out of 17 patients were noted but in 27 patients with 11–15 years of duration only 1(3.7037%) was diagnosed as moderate epithelial dysplasia (100%).

In CS only 1 patient with severe epithelial dysplasia was noted out of 2 patients in 11–15 years of duration.

In TA 5 out of 5 patients (100%) were diagnosed as reactive hyperplasia with Dysplastic changes in overlying epithelium (1) (20%), moderate epithelial dysplasia (2) (40%), severe epithelial dysplasia (1) (20%) and carcinoma in situ in 1 (20%).

In case of C with 0–5 years of duration 4 out 5 patients of OSMF were diagnosed as very early OSMF (2) (50%) and early OSMF (2) (50%). With 6–10 years of duration 3 patients with early OSMF (60%) and 2 with advanced OSMF (40%) out of 6 patients were diagnosed. In 8 patients with 11–15 years

Table 3.5 Correaltion of Habit, Duration, Clinical Preamalignant Disorder (250) with Histopathology (50)

Habit	Duration (in years)	Leukoplakia		OSMF		Erythroplakia		Verrucous Lesion	
		Total No.(Out of 250)	Histopathological Evalution (Out of 50)	Total No. (Out of 250)	Histopathological Evalution (Out of 50)	Total no.(Out of 250)	Histopathological Evalution (Out of 50)	Total No. (Out of 250)	Histopathological Evalution (Out of 50)
Chewers	0–5	67	11 (16.42%) Mild ED = 6 (54.55%) Moderate ED = 4 (66.67%) Hyper-orthrokeratinized epithelium = 1 (9.10%)	5	4 Very early OSMF = 2(50%) Early OSMF = 2(50%)	0	0	0	0
	6–10	60	9 Mild ED = 1(11.11%) Moderate ED = 5(55.56%) Severe ED = 1(11.11%) Carcinoma in situ = 1 (11.11%) Atrophic epithelium with inflamed connective tissue = 1 (11.11%)	6	5 Early OSMF = 3(60%) Advanced OSMF = 2 (40%)	0	0	0	0
	11–15	14	1 Carcinoma in situ = 1 (100%)	8	6 Advanced OSMF = 1 (16.67%) Moderately Advanced OSMF = 2(33.33%) OSMF with ED = 2 (33.33%) OSMF with PVL = 1 (16.67%)	1	0	3	1 Verrucous Carcinoma = 1 (100%)
	16–20	0	0	0	0	0	0	0	0

Table 3.5 *Continued*

Smokers	0–5	10	2 Epithelial Hyperplasia = 1(50%) Hyper-orthrokeratinization = 1(50%)	0	0	0	0
	6–10	17	2 Acanthosis = 1(50%) Hyper-parakeratinization with Acanthosis = 1(50%)	0	0	0	0
	11–15	27	1 Moderate ED = 1(100%)	0	0	0	0
	16–20	0	0	0	0	0	0
Chewers + smokers	11–15	2	1 Severe ED = 1(100%)	0	0	0	0
Tobacco + alcohol	11–15	5	5 Reactive hyperplasia with Dysplastic changes in overlying epithelium = 1(20%) Moderate ED = 2(40%) Severe ED = 1(20%) Carcinoma in situ(20%)	2 OSMF with SCC = 2 (100%)	0	0	0

of duration 6 were diagnosed as advanced OSMF (1) (16.67%), moderately advanced OSMF (2) (33.33%), (2) OSMF with epithelial dysplasia (33.33%) and (1) with OSMF with proliferative verrucous leukoplakia (16.67%).

The TA were 2 out of 2 diagnosed as OSMF with SCC.With 11–15 years of duration of chewing 1 diagnose as verrucous carcinoma out of 3 patients (Table 3.5).

Paired Z TEST

By applying Z test for proportion to test significant difference between Leukoplakia and OSMF for different duration in Chewer + Smoker, Alcohol and Tobacco + Alcohol a significant difference was observed from 11–15 and 16–20 years of duration i.e $P < 0.05$

For Smoker						
S. No.	Source of Varitation		S.S	M.S.S	Fratio	Ftab
1.	Due to Precancerous Disorder	1	242	242	7.356	5.99
2.	Error/Residual	6	197.3898	32.8983	—	
Total		7	439.3898	—	- -	
$P < 0.05\%$						

For Chewers						
S. No.	Source of Variation		S.S	M.S.S	Fratio	Ftab
1.	Due to Precancerous Disorder	3	3382	1127.33	6.0588	4.49
2.	Error/Residual	12	2232.78	186.0649	—	
Total		15	5614.78	—	—	—
$P < 0.05\%$						

One Way Anova – F Test

To test the significant difference among the Leukoplakia, OSMF, Erythroplakia and Verrucous lesion One Way ANOVA – F test was applied at 5% level of significance.

A significant difference was found among the clinical lesions at 5% level of significance with duration of time.

A significant difference was observed for chewers and smokers i.e $P < 0.05$.

4

Discussion

Over 90% of malignancies of the oral cavity are squamous cell carcinoma arising from the lining mucosa and the vast majority of oral squamous cell carcinomas are preceded by precursor lesions that can present as either leukoplakia, erythroplakia, oral submucous fibrosis or verrucous lesions. As an early sign of damage to the oral mucosa, individuals often develop clinically visible whitish (leukoplakia) or reddish (erythroplakia) lesions and oral submucous fibrosis (OSF). All these well-established precancerous lesions are easily diagnosed and present an important indicator of oral cancer risk. The risk of developing oral cancer was found to become higher with increasing duration of deleterious habits per day.

In India, oral squamous cell carcinoma constitute a serious problem thus a study of oral precancer is of great importance. A convincing correlation has been established between oral cancer and tobacco usage in the form of chewing (pan masala, gutkha, khaini, pattiwala tobacco, mainpuri tobacco, mishri, zarda, kiwam, gudhakhu) and smoking (bidi, cigarette–filtered/unfiltered, cigars-chutta (reverse smoking), pipe–chillum or sulpa, water pipe-hookah) as well as between various precancerous disorders and similar forms of tobacco usage. The use of alcohol drinking associated with the habit of tobacco usage was also appear to increase the risk of oral precancerous disorder by several ways either directly or indirectly. Alcohol–specifically ethanol is the most socially accepted addictive drug which can have serious health hazards.

Since previous studies done by various authors **J.J. Pindborg (1963–64)**,[22] **P.N. Wahi (1964–66)**[23] reported that the habits of tobacco usage and alcohol drinking was prevalent in Western Uttar Pradesh, but no study of such nature was conducted in Moradabad (Western Uttar Pradesh) so we have conducted survey study of habit positive individuals in Moradabad to find out the Epidemiological, Clinical and Histopathological correlation of premalignant disorders which include leukoplakia, OSMF, Erythroplakia and verrucous lesions with Tobacco users and Alcohol drinkers. The present study was inspired and design according to studies done by various authors like **Jerry E. Bouquot and Robert J. Gorlin** in **1957–1973** (23,616),[27] **P.N. Wahi March 1964** and **September 1966** (126,063),[23] **Jens J. Pindborg et al.** in **1967**, (10,000),[22] **Fali S Mehta et al.** in **1969** (50,915),[75] **Harry H. Mincer** and **Kenneth P. Hopkins 1972** (50,000),[13] **Reichart PA 1979–1984** (1866),[33] **PC Gupta** 1980 (50,915),[76] **Khin Maung Lay et al.** in **1982**, (6000 villagers),[34] **Prakash C. Gupta 1984** (10,914),[21] **John F. Wisniewski** and **Alfred A. Bartolucci 1987** (858),[36] **Tang J-G et al. 1997** (100 000),[52] **A. Ariyawardana et al. 1999** (12,716),[9] **Mia Hashibe, et al.** in **2000** (47,773),[56] **D.N. Sinha 2001** (15,247),[17] **V.K. Hazarey et al.** in **2007** (1000).[73]

The presence of 250 clinically lesion positive in 35,000 habit positive individuals was comparable with other studies like (**Mia Hashibe** (100/47,773),[56] **P.N. Wahi** (600/126,063),[23] **J.J Pindborg** (509/10,000),[22] **Bouquot et al.** (798/23,616),[27] **Fali. S Mehta** (723/50,915),[75] **Harry H. Mincer** and **Kenneth P. Hopkins** (67/50,000),[13] **Khin Maung Lay et al.** (295/6000 villagers).[34]

In our study male predominance was found which equivocal with other studies (**Jerry E. Bouquot and Robert J Gorlin 1957–73**,[27] **J.J. Pindborg 1963–64**,[22] **Prakash C Gupta 1984**,[21] **Jolan Banoczy et al. 2001**,[57] **D.N Sinha 2003**,[16] **Jyotsna Changrani et al. 2003–2004**).[70] The habit of tobacco with alcohol were more commonly seen in age group of 26–35 years however lesions with tobacco chewing were more commonly seen in 36–45 years which were similar to the fact that the prevalence rate for oral cancer increases with advancing age must be a reflection of the effect of a latency period intervening after the start of a chewing habit in younger persons, in addition to the higher susceptibility to cancer of older persons.

In the present study the premalignant disorders were more prevalent in chewers and the findings of the present study were correlating with the studies done by **Fali. S. Mehta (1972)**,[29] **Curtis J. Creath (1991)**,[44] **S. Warnakulasuriya (2004)**,[77] **Urmila Niar (2004)**.[63] They showed a causal association between tobacco and betel quid chewing habits and oral mucosal diseases such as leukoplakia, oral submucous fibrosis and heavy users have a significantly increased mortality rate. But some of the studies by, **J.E. Henningfield (1997)**,[78] **Deborah M. (2008)**[79] also showed that Tobacco smoking is an important risk factor for precancerous lesions of the mouth. Smokers have a significantly higher prevalence of leukoplakia compared with non-smokers. It has also been demonstrated that there is a dose-response relationship for tobacco use and the risk of malignant transformation of oral leukoplakia. The presence of leukoplakia increased with the number of smoked cigarettes per day. The majority of smokers with leukoplakia (74.0%) smoked more than 20 cigarettes per day compared to 34.5% of those without leukoplakia.

In the our study Leukoplakia was the most common precancerous lesion and erythroplakia was least common in relation to habits which was equivocal with the findings of various authors. {**Fali S Mehta (1972),**[29] **Teruo Amagasa,**[80] **Marcio Diniz Freitas (2006)**}.[71]

The present study showed a positive correlation between the duration and associated precancerous disorders, it was proved by **One Way Anova analysis and Paired Z test (P < 0.05%)**. It was shown that as the duration of habit increased, the prevalence of dysplastic and carcinomatous changes in precancerous disorders were also increased. This finding of the present study was related to the study done by **(Wahi, 1964–66,**[23] **Macigo et al. 1996).**[50] They showed that the longer the duration of quid chewing, the higher the rate of oral cancer and a clear cut tendency was noted that the earlier the habit of chewing started, the higher the risk of developing oral cancer.

In chewers the histopathologically diagnosed lesions were mild epithelial dysplasia, moderate epithelial dysplasia, severe epithelial dysplasia, carcinoma in situ, less frequently hyperorthokeratinization, very early, early, moderately advanced and advanced OSMF, PVL and verrucous carcinoma. These lesions were correlated with the duration of the habit such as: in small duration mild epithelial dysplasia were found more and as the duration of habit increased the grades of dysplasia were also increased. In case of OSMF, very early OSMF were found more and as the duration of habit increased the grades of dysplasia were also increased. Along with the different stages of OSMF less frequently proliferative verrucous leukoplakia was also reported in relation to OSMF. One case of verrucous carcinoma was also reported as the duration of the habit increased. These findings of the present study were correlated with the findings of **Fali S Mehta (1969),**[81] **Wahi (1964–66).**[23]

In smokers the histopathologically diagnosed lesions were Epithelial hyperplasia, Hyperorthokeratinization, Acanthosis, moderate epithelial dysplasia while in smokers as well as chewers the lesions were mild, moderate, severe Epithelial dysplasia, carcinoma in situ and OSMF with S.C.C. So in the present study it was shown that the dysplastic lesions with carcinomatous changes were found to be more common among the chewers. This finding of the present study was correlated to the findings of the study done by **Jolan Banoczy (2001),**[57] **Marija Bokor-Bratic (2002),**[59] **Douglas E. Morse (2007)**[82] who stated that both oral cancer and oral leukoplakia can be induced and promoted by tobacco. In tobacco users with alcohol drinkers the histopathologically diagnosed lesions were epithelial dysplasia and OSMF with S.C.C. This finding was also showed in the study of previous authors (**Ko YC, Huang, K. Andre 1995**).[49] They suggested that alcohol consumption alone was not independently associated with oral cancer, it did seem to enhance the risk of developing the oral cancer when used in combination with the other two habits.

Oral health is strongly related to various habits which are highest among the people of low socioeconomic background. The numerous harmful effects of these habits usage on oral health range from stained teeth, halitosis, and precancerous mucosal disorders to fatal oral cancer. As members of an important health profession, this is our duty to promote oral health and healthy lifestyles among our patients, by raising their awareness about the harmful effects of tobacco and alcohol on health and guiding them in conquering these deleterious addictions.

5

Summary

The Endeavour of present survey study in Moradabad of Oral Precancerous Disorders in Tobacco users and Alcohol Drinkers with Special Emphasis on Epidemiology, Clinical and Histopathological Correlation with Habits.

The aims and objectives of the study were to investigate the prevalence of premalignant disorders in Moradabad population caused in tobacco users, alcohol drinkers and gave special emphasis on:

1. Epidemiology correlation
2. Clinical correlation and
3. Histopathological correlation

All the patients reported to OPD of Kothiwal Dental College and Research centre were questioned for the type and nature of oral habits by the BDS final year students posted in the Department of Oral Medicine irrespective of chief complaint. The habit positive patients were segregated and further examined for the presence of any oral lesions by posted interns in the same department. After that the clinically positive premalignant disorder cases were further examined by me under the supervision of faculty members in Department of Oral Pathology and Microbiology. After taking written consent from the patients the biopsy was performed and the tissues were processed, sections obtained and stained with H&E for their histopathological correlation. After completing all the Aims and Objectives the following results were achieved:

1) The total survey of 35,000 habit positive individuals was done in which 250 patients were identified of different premalignant disorders such as Leukoplakia, OSMF, Erythroplakia, and Verrucous lesions were found.
2) The age group involved in habit positive individual 26–35 and in lesion positive individuals was 36–45 years with male predominance.
3) In clinical positive lesions Leukoplakia was commonest and Erythroplakia was least common.

4) Out of 250 habit positive individuals tobacco chewing was the commonest habit and only 50 patients were readily diagnosed histopathologically.

5) Histopathologically Epithelial dysplasia was commonly found lesion associated with habit. Dysplasia was increased in the lesions as the duration of the habit was increased from 0–5 years to 11–16 years with the presence of carcinomatous changes in the lesion.

6

Conclusion

The present survey study was conducted in Moradabad population to find out the premalignant disorder in Tobacco users and Alcohol drinkers with special emphasis given on epidemiological, clinical, histopathological correlation. The study was done in Out Patient Department (OPD) of Kothiwal Dental College and Research Centre in Moradabad and Oral cancer detection camps held in and around the Moradabad population. From the foresaid study the following conclusions are obtained:

(1) The most commonly affected age was third and fourth decade of life. Males were seen to be more commonly affected than the females.

(2) Premalignant disorders were closely related to the deleterious habits such as tobacco chewing. Tobacco and alcohol drinking together was also considered as a risk factor for such lesions.

(3) As the duration of habits increased, the precancerous disorders became more dysplastic.

To conclude a positive correlation was found between habits and precancerous disorders and it was noticed that as the duration of the lesion increased, the lesion became more Dysplastic with carcinomatous changes. So a thorough knowledge of these deleterious habits in precancerous lesions and conditions in the oral cavity is of such paramount importance to the dentist is linked with the central position the dentist holds as regards the early-diagnosis and management of oral cancer.

Bibliography

[1] P.M. Speight, P.M. Farthing and J.E. Bouquot; The Pathology of Oral Cancer And Precancer; Current Diagnostic Pathology (1996), 3, 165–176

[2] T. Axell, J.J. Pindborg, C.J. Smith, I. Van Der Waal and an International Collaborative Group on Oral White Lesions; Oral White Lesions with Special Reference to Precancerous and tobacco related Lesions: conclusions of an international symposium held in Uppsala, Sweden, May 18–21, 1994, *J Oral Pathol Med* 1996; 25; 49–54

[3] J. Banoczy and L. Sugar; Longitudinal studies in Oral Leukoplakias; *J. Oral Path*, 1972: 1: 265–272

[4] S.R. Prabhu, D.F. Wilson, D.K. Daftary and N.W. Johnson; Oral Diseases in the tropics; 2003; 105–108

[5] F.G. Macrgo, D.L. Mwaniki, S.W. Guthua, The association between oral leukoplakia and use of tobacco, alcohol and khat based on relative assessment in Kenya; *Eur J Oral Sci*; 1995; 103; 268–273

[6] WHO collaborating centre for Oral Precancerous lesions; Definition of leukoplakia and related lesions: An Aid to studies on oral precancer; 46, 4, *Oral Surg*, October 1978, 518–539

[7] Peter A. Reichart, Hans Peter Philipsen; Oral erythroplakia—a review; Oral Oncology (2005) 41, 551–561

[8] K. Ranganathan et al. Oral sub mucous fibrosis; a case control studying Chennai, South India; *J Oral Pathol Med* (2004) 33: 274–7

[9] A. Ariyawardana et al. Effect of betel chewing, tobacco smoking and alcohol consumption on oral sub mucous fibrosis: a case–control study in Sri lanka; *J Oral Pathol Med* (2006) 35: 197–201

[10] P.N. Sinor, P.C. Gupta, P.R. Murti, R.B. Bhonsle, D.K. Daftary, E.S. Mehta, J.J. Pindborg: A case-control study of oral submucous fibrosis with special reference to the etiologic role of areca nut, *J Oral Pathol Med* 1990; 19: 94–8

[11] W.M. Tilakaratne, M.F. Klinikowski, Takashi Saku, T.J. Peters, Saman Warnakulasuriya; Oral submucous fibrosis: Review on aetiology and pathogenesis; Oral Oncology (2006) 42, 561–568

[12] Selwyn Jacobson And Mervyn Shear; Verrucous carcinoma of the mouth; *J. Oral Path.* 1972: 1: 66–75

[13] Harry H. Mincer, Sidney A. Coleman and Kenneth P.Hopkins; Observations on the clinical characteristics of oral lesions showing histologic epithelial dysplasia; *Oral Surg*; March 1972; 33; 3; 389–399

[14] S. Warnakulasuriya, Newell. W. Johnson, I. van der Waal; Nomenclature and classification of potentially malignant disorders of the oral mucosa–Review; *J Oral Pathol Med* (2007)

[15] Ching–Hung Chung et al. Oral precancerous disorders associated with areca quid chewing, smoking and alcohol drinking in Southern Taiwan; *J Oral Pathol Med* (2005) 34: 460–6

[16] Dhirendra N. Sinha, Prakrash C. Gupta, Mangesh S. Pednekar; Tobacco use in a rural area of bihar, India; *Indian journal of community medicine*, Vol. XXVIII, no. 4, Oct–Dec, 2003, 167–170

[17] D. N. Sinha; Gutka Advertisement and smokeless tobacco use by adolescents in Sikkim, India; *Indian Journal Of community Medicine*; Vol. 30, No. 1, Januray–March, 2005

[18] C. Sushma, C. Sharang: Pan Masala advertisement are surrogate for tobacco products; Indian Journal of Cancer; April–June 2005, Vol 42, Issue 2; 94–98

[19] K. Kiran Kumar, T.R. Saraswathi, K. Ranganathan, M. Uma Devi, Joshua Elizabeth; Oral Sub mucous fibrosis: Aclinco-histopathological study in Chennai: *Indian J Dent Res*, 18 (3), 2007; 106–111

[20] P. Rajalalitha, S. Vali; Molecular pathogenesis of Oral Sub Mucous Fibrosis–a collagen metabolic disorder; *J Oral Pathol Med* (2005) 34: 321–8

[21] Prakash C. Gupta: Epidemiologic study of the association between alcohol habits and oral leukoplakia; *Community Dent Oral Epidemiol* 1984; 12: 47–50

[22] J.J. Pindborg, Joyce Kiaer, P.C. Gupta, T.N. Chawla; Studies in Oral Leukoplakias–Prevalence of Leukoplakia among 10,000 Persons in Lucknow, India with Special Reference to Use of Tobacco and Betel Nut; *Bull. Wld Hlth Org*; 1967; 37; 109–116

[23] P.N. Wahi: The Epiemiology of Oral and Oropharyngeal Cancer–A Report of the study in Mainpuri District, Uttar Pradesh, India; *Bull Wld Hlth Org*, 1968, 38, 495–521

[24] Jolan Bancozy; Follow-up Studies in Oral Leukoplakia; *J. max.-fac. Surg.* 5 (1977) 69–75

[25] William G. Shafer; Oral carcinoma in situ; *Oral Surg*, Febraury, 1975. vol. 39; 2; 227–238

[26] Renstrup, Grete (1958) 'Leukoplakia of the Oral Cavity: A Clinical and Histopathologic Study', Acta Odontologica Scandinavica, 16:1, 99–111

[27] Jerry E. Bouquot, and Robert J. Gorlin; Leukoplakia, lichen planus, and other oral keratoses in 23,616 white Americans over the age of 35 years; *Oral Surg. Oral Med. Oral Pathol.* 61: 373–381, 1986

[28] Charles A. Waldron and William G Shafer; Leukoplakia Revisited–A Clinicopathologic Study 3256 Oral Leukoplakias; Cancer 36: 1386–1392, 1975

[29] Fali S. Mehta, B.C. Shroff, P.C. Gupta, and D.K. Daftary, Bombay, India; Oral leukoplakia in relation to tobacco Habits-A ten-year follow-up study of Bombay policemen; *Oral Surg*, Volume 34, Number 3, September, 1972, 426–433

[30] Jolan Bancozy and Arpad Csiba; Occurrence of epithelial dysplasia in oral leukoplakia-Analysis and follow-up study of 12 cases; Volume 12 Number 6; *Oral Surg*, December, 1976, 766–774

[31] Jens J. Pindborg, D.D.S., Dr.Odont., Odont.Dr.h.c.,Dinesh K. Daftary, B.D.X., M.D.S., and Fali S. Mehta, B.D.S., D.M.D., M.S.; A follow-up study of sixty-one oral dysplastic precancerous lesions in Indian villagers; *Oral Surg*, March, 1977; Volume 43; Number 3, 383–390

[32] Fali S. Mehta, P.N. Jalnawalla, D.K. Daftary, P.C. Gupta and J.J. Pindborg; Reverse smoking in Andhra Pradesh, India: Variability of clinical and histologic appearances of palatal changes; *Int. J. Oral Surg.* 1977: 6: 75–83

[33] P.A. Reichart, U. Mohr, S. Srisuwan, H. Geerlings, C. Theetranont, T. Kangwanpong; Precancerous and other oral mucosal lesions related to chewing, smoking and drinking habits in Thailand, Community Dent Oral Epidemiol 1987; 15; 152–60

[34] Lay, K, M,, Sein, K,, Myint, A,, Ko, S, K, & Pindborg, J, J,: Epidemiologic study of 6000 villagers of oral precaneerous lesions in Bilugyun: preliminary report, Gommunity Dent, Oral Epidemiol, 1982: 10: 152–155

[35] P.R. Murti, R.B. Bhonsle, J.J. Pindborg, D.K. Daftary, P.C. Gupta, F.S. Mehta; Malignant transformation rate in oral submucous fibrosis over a 17-year period. *Community Dent Oral Epidemiol* 1985; 13; 340–1

[36] J.F. Wisniewski, A.A. Bartolucci; Comparative patterns of smokeless tobacco usage among major league baseball personnel. *J Oral Pathol Med* 1989; 18; 322–326

[37] M.D. Wolfe, J.P. Carlos: Oral health effects of smokeless tobacco use in Navajo Indian adolescents. *Community Dent Oral Epidemiol* 1987; 15: 230–5

[38] Satyanarayana gavarasana and Maha deva sastri susarla; palatal mucosal changes among reverse smokers in an Indian villages; *Jpn, J.Cancer Res* 80 209–211, March 1989

[39] Satyanarayana gavarasana, Vijaya Prasad Doddi, Gorty VSNR Prasad, Apparao Allam and Bellana S.R. Murthy; A smoking survey of college students in India: Implication for designing an Antismoking Policy; *J Cancer Res*; 82, 142–145, Feb. 1991

[40] Gupta PC: Leukoplakia and incidence of oral cancer. *J Oral Pathol Med* 1989; 18: 17

[41] S. Mehta Fali, R.B. Bhonsle, P.R. Murti, D.K. Daftary, P.C. Gupta, J.J. Pindborg: Central papillary atrophy of the tongue among bidi smokers in India; a 10-year study of 182 lesions. *J Oral Pathol Med* 1989; 18: 475–480

[42] P.E. Petersen: Smoking, alcohol consumption and dental health behavior among 25–44-year-old Danes. *Scand J Dent Res* 1989; 97: 422–31

[43] P.R. Murti, P.C. Gupta, R.B. Bhonsle, D.K. Daftary, F.S. Mehta, J.J. Pindborg: Effect on the incidence of oral submucous fibrosis of intervention in the areca nut chewing habit. *J Oral Pathol Med* 1990; 19: 99–100

[44] Curtis J. Creath, DMD, MS, a Gary Cutter, PhD,b Dorothy H. Bradley, MSc and J. Timothy Wright, DDS, MSd Birmingham, Ala.; Oral leukoplakia and adolescent smokeless tobacco use; *Oral Surg Oral Med Oral Pathol* 1991; 72: 35–41

[45] Tatiana V. Evstifeeva and David G. Zaridze; Nass use, Cigarette Smoking, Alcohol Consumption and Risk of Oral and Oesophageal Precancer; Oral Oncology, *Eur J Cancer*, Vol. 28B, No. 1, pp. 29–35, 1992

[46] Ko YC, T.A. Chiang, S.J. Chang, S.F. Hsieh: Prevalence of betel quid chewing habit in Taiwan and related sociodemographic faetors. *J Oral Pathol Med* 1992: 21: 261–264

[47] Maher R. Lee A.1, Warnakulasuriya KAAS, J.A. Lewis, N.W. Johnson: Role of areca nut in the causation of oral submucous fibrosis: a case-control study in Pakistan. *J Oral Pathol Med* 1994: 23: 65–69

[48] Reichart, Peter A: Oral cancer and precancer related to betel and miang chewing in Thailand: a review. *J Oral Pathol Med* 1995; 24: 241–243

[49] Y.C. Ko, Y.L. Huatig, C.H. Lee, M.J. Chen, L.M. Litt, C.C. Tsai: Betel quid chewing, cigarette smoking and alcohol consumption related to oral cancer in Taiwan. *J Oral Pathol Med* 1995; 24: 450–3

[50] Francis Githua Macigo, David Lemmy Mwaniki and Symon Wangombe Guthua; Influence of dose and cessation of kiraiku, cigarettes and alcohol use on the risk of developing oral leukoplakia; *Eur J Oral Sci*; 1996; 104; 498–502

[51] T. Nagao, N. Ikeda, H. Fukano, H. Miyazaki, M. Yano, S. Warnakulasuriya; Outcome following a population screening programme for oral cancer and precancer in Japan; Oral Oncology 36 (2000) 340–346

[52] J.G. Tang, X.F. Jian, M.L. Gao, T.Y. Ling, K.H. Zhang: An epidemiological survey of oral submucous fibrosis in Xiangtan City, Hunan Province, China. *Cotninunity Dent Oral Fpidemiol* 1997; 25: 177–80

[53] R.B. Zain, N. Ikeda, I.A. Razak, T. Axell, Z.A. Majid, P.C. Gupta, M. Yaacob: A national epidemiological survey of oral mucosal lesions in Malaysia. *Community Dent Oral Fpidemiol* 1997; 25; 377–83

[54] N. Shah, P.P. Sharma: Role of chewing and smoking habits in the etiology of oral submucous fibrosis (OSF): a case-control study. *J Oral Pathol Med* 1998; 27: 475–9

[55] W.F. Allard, E.B. DeVol, O.B. Te: Smokeless tobacco (shamma) and oral cancer in Saudi Arabia. *Community Dent Oral Epidemiol* 1999; 27: 398–405

[56] Mia Hashibe, Babu Mathew, Binu Kuruvilla,Gigi Thomas, Rengaswamy Sankaranarayanan, Donald Maxwell Parkin, and Zuo-Feng Zhang; Chewing Tobacco, Alcohol, and the Risk of Erythroplakia; Cancer Epidemiology, Biomarkers & Prevention Vol. 9, 639–645, July 2000

[57] Jolán Bánóczy, Zeno Gintner, Csaba Dombi; Tobacco Use and Oral Leukoplakia; Journal of Dental Education, April 2001, Volume 65, No. 4

[58] Yi-Hsin Yang, Hsiu-Yu Lee, Sen Tung, Tien-Yu Shieh; Epidemiological survey of oral submucous fibrosis and leukoplakia in aborigines of Taiwan; *J Oral Pathol Med* 2001: 30: 213–9

[59] Marija Bokor-Bratic, Nada Vuèkovic; Cigarette smoking as a risk factor associated with oral leukoplakia; Archive of Oncology 2002; 10(2): 67–70

[60] Rodu B, Stegmayr B, Nasic S, Asplund K; Impact ofsmokeless tobacco use on smoking in northern Sweden. *J Intern Med* 2002; 252: 398–404

[61] Sinha DN, Gupta PC, Pednekar MS; Tobacco use among students in the eight Northeastern states of India; Indian *J Cancer* 2003; 40: 43–59

[62] M. Hashibe, B.J. Jacob, G. Thomas, K. Ramadas, B. Mathew, R. Sankaranarayanan, Z.F. Zhang; Socioeconomic status, lifestyle factors and oral premalignant lesions; Oral Oncology (2003) 39 664–671

[63] Urmila Nair, Helmut Bartsch and Jagadeesan Nair; Alert for an epidemic of oral cancer due to use of the betel quid substitutes gutkha and pan masala: a review of agents and causative mechanisms; *Mutagenesis* vol. 19 no. 4, pp. 251–262, 2004

[64] S.V. Subramanian, Shailen Nandy, Michelle Kelly, Dave Gordon, George Davey Smith; Patterns and distribution of tobacco consumption in India: cross sectional multilevel evidence from the 1998–9 national family health survey; BMJ 2004; 328: 801–6

[65] Sharma Rameshwar, Pednekar Mangesh S, Rehman AU, Gupta Rakesh; Tobacco use among school personnel in Rajasthan, India; *Indian J Cancer* 2004; 41: 162–6

[66] Y.H. Yang, C.H. Chen, J.S.F. Chang, C.C. Lin, T.C. Cheng, T.Y. Shieh; Incidence rates of oral cancer and oral pre-cancercous lesions in a 6-year follow-up study of a Taiwanese aboriginal community; *J Oral Pathol Med* (2005) 34: 596–601

[67] Y.H. Yang, Y.C. Lien, P.S. Ho, C.H. Chen, J.S.F. Chang, T.C Cheng, T.Y. Shieh; The effects of chewing areca/betel quid with and without cigarette smoking on oral submucous fibrosis and oral mucosal lesions; Oral Diseases (2005) 11, 88–94

[68] M.A. Fisher, J.E. Bouquot, B.J. Shelton. Assessment of risk factors for oral leukoplakia in West Virginia. *Community Dent Oral Epidemiol* 2005; 33: 45–52

[69] Glorian Sorensen, Prakash C. Gupta, and Mangesh S. Pednekar; Social Disparities in Tobacco Use in Mumbai, India: The Roles of Occupation, Education, and Gender; *Am J Public Health*. 2005; 95: 1003–1008, doi:10.2105/AJPH.2004.045039

[70] Jyotsna Changrani, Francesca M. Gany, Gustavo Cruz, Ross Kerr, Ralph Katz; Paan and Gutka Use in the United States: A Pilot Study in Bangladeshi and Indian-Gujarati Immigrants in New York City; *J Immigr Refug Stud*. 2006; 4(1): 99–110

[71] Marcio Diniz Freitas, Andrés Blanco-Carrión, Pilar Gándara-Vila, José Antúnez-López, Abel García-García and José Manuel Gándara Rey, Santiago de Compostela, spain Universidad De Santiago De Compostela; Clinicopathologic aspects of oral leukoplakia in smokers and Nonsmokers, *Oral Surg Oral Med Oral Pathol Oral Radiol Endod* 2006; 102: 199–203

[72] Sardar Z Imam, Haq Nawaz, Yasir J Sepah, Aqueel H Pabaney,Mahwish Ilyas and Shehzad Ghaffar; Use of smokeless tobacco among groups of Pakistani medical students–a cross sectional study; BMC Public Health 2007, 7: 231

[73] V.K. Hazarey, D.M. Erlewad, K.A. Mundhe, S.N. Ughade; Oral submucous fibrosis: study of 1000 cases from central India; *J Oral Pathol Med* (2007) 36: 12–17

[74] Fali S. Mehta, J.J. Pindborg, P.C. Gupta and D.K. Daftary; Epidemiologic and Histologic study of Oral Cancer and Leukoplakia among 50, 915 villagers in India; Cancer October 1969; 4; 24; 832–849

[75] P.C. Gupta, Fali S Mehta, D.K. Daftary, J.J. Pindborg, R.B. Bhonsle, P.N. Jalnawala, P.N. Sinor, V.K. Pitkar, P.R. Murti, R.R. Irani, H.T. Shah, P.M. Kadam, K.S.S. Iyer, H.M. Iyer, A.K. Hedge, G.K. Chandrashekhar, B.E. Sahiar and M.N. Mehta; Incidence rates of Oral Cancer and natural history of Oral Precancerous lesions in a 10-year follow up study of Indian Villagers; *Community Dent. Oral Epidemiol*, 1980; 8; 287–333

[76] S Warnakulasuriya; Smokeless Tobacco and Oral Cancer; Oral Diseases (2004); 10; 1–4

[77] J.E. Henningfield, R.V. Fant, S.L. Tomar; Smokeless Tobacco: An Addicting Drug; *Adv Dent Res*, September, 1997; 11(3), 330–335

[78] Deborah M. Winn; Smokeless Tobacco and Cancer: The Epidemiologic Evidence; *CA Cancer J Clin*; 1988; 38; 236–243

[79] Teruo Amagasa, Masashi Yamashiro and Hitoshi Ishikawa; Oral Leukoplakia Related to Malignant Transformation; Oral Science International; November 2006; vol. 3, no. 2; 45–55

[80] Fali S Mehta, D.K. Daftary, B.C. Shroff and L.D. Sanghvi; Clinical and Histologic study of oral leukoplakia in relation to habits; vol. 28, no. 3, OS, OM & OP, September 1969, 372–388

[81] Douglas E. Morse, Walter J. Psoter, Deborah Cleveland, Donald Cohen, Miresyed Mohit–Tabatabai, Diane L.Kosis and Ellen Eisenberg; Smoking and drinking in relation to oral cancer and oral epithelial dysplasia; Cancer Causes Control, 2007 November; 18(9); 919–929

About the Author

Dr Manjul Tiwari, born on 13 Jan 1980 is currently working as Senior Lecturer, School of Dental Sciences, Sharda University, India from August 2009 after pursuing MDS from Kothiwal Dental College Moradabad in Oral Pathology and Microbiology.

He has over 40 publication on various fields of dentistry, genetics, forensic odontology in highly impact factor national and international journals ranging from Cancer therapies, Tumor Markers, gene therapy, child abuse, nanotechnology and forensic dentistry which has been published in Journal of Cancer Research and Therapeutics, Journal of Natural Science, Biology & Medicine, Journal of Oral and Maxillofacial Pathology, Indian Journal of Human Genetics, Guident, IDA Times, etc. He was invited Guest speaker in Hohhot, Inner Mongolia, China, Shanghai, Montreal (Canada) etc where he gave Oral Presentation on Tumor Immunology and Cancer Immunotherapy. Currently he is Reviewer in various journals like Journal of Oral and Maxillofacial Pathology.

He has collaborated Memorandum of Understanding with School of Bioengineering, McGill University, Canada and State University of New York at Buffalo, (Buffalo University), USA.

He has attended various CDE's and workshops all over India. He has presented First and Second Prize winning Papers and Poster on "DNA Profiling or Fingerprinting" in XIX National & First International Conference of IAOMP, Radisson Temple Bay, Chennai, "Bite Marks: Role of Saliva in Human Genome" in XV National conference of IAOMP, Chennai, "Influence of inflammation on the polarization colors of collagen fibers in the wall of Odontogenic Keratocyst and their Clinico–Pathological Correlation" in VIII National Post Graduate Convention of IAOMP, Wardha and "Comparative analysis of primary intraosseous carcinoma & ameloblastic carcinoma: Case Reports with review of literature" in XVI National Conference of IAOMP, Khajuraho.

He is life member of several presitigious associations like Indian Dental Association (IDA), Indian Association of Oral and Maxillofacial Pathology

(IAOMP), International Association of Oral Pathologist (IAOP) and International Association of General Dentistry (IAGD).

He has done various industrial training Program like on Nanotechnology, Biotechnology, Cancer Genetics etc, Advanced Fixed Prosthodontics, Smile design as well as he hold PG Diploma in Hospital Administration. His First national book on Modern Dictionary of Human anatomy from Deep & Deep Publications published last year and this book is his first International publication.

Printed and bound by CPI Group (UK) Ltd, Croydon, CR0 4YY

23/10/2024

01777667-0019